I0469321

Waves

1.0 WAVES

| 1 |

Nature makes everything unique and simple in that it has helped us to understand the environment around us; waves are other important aspects of physics that helps mankind to appreciate the usefulness of nature. Your ability to hear your teacher's speech is due to waves, tuning your transistor radio to a particular frequency and receiving signals is due to waves. Telecommunication, satellite communication, oil and gas exploration, internet accessibility and lots more are feasible as a result of waves. In this chapter, we are going to "x-ray" the hidden facts applicable to physical life.

Just relax and get the gist!!!

1.1 Definition of Waves

| 2 |

Waves can occur whenever a system is disturbed from equilibrium and the disturbance carries energy as it travels from one region to another. Waves are generated from a source and they transfer energy as they travel. Some waves requires material medium for propagation while some does not. Waves are generated set of pulses, from a source, which transfer energy from one point to another as they travel.

A wave is a disturbance which travels through a medium and transfers energy from one point to another without causing any permanent displacement of the medium itself.

Examples are ripples on water, set of pulse generated from a plucked string, musical sounds, seismic tremor triggered by an earth quake, etc.

Yes! Very important to remember about waves is that energy is transferred, but the medium through which they travel is not permanently displaced. This is why water flow is different from a water wave; river flows are not water waves!!! Not at all.

1.2 Wave motion, or wave front wave forms

3

Wave forms are the shapes of the waves and the shape of the waves depends on the nature of the medium. Wave fronts are lines or positions taken by an advancing (or progressive) wave in which all the particles are in the same phase. A Wave motion is a process of transferring energy from one point of the medium to another. As waves transport energy, the matter or particles of the medium do not move but only vibrate about their mean or rest position.

1.3 Classification of waves

4

Waves as a mode of energy transfer are generated from a source and the wave motion occurs immediately when there is disturbance in the medium. Waves are classified into

(i)   Mechanical wave
(ii)   Electromagnetic waves and

Mechanical waves are waves that require material medium for their propagation. Mechanical waves are governed by Newton's laws and they can exist only within a material medium, like air, rock, water, metals, plastic, woods, etc. Examples of mechanical waves are water waves, sound waves, seismic waves, waves generated from a plucked rope or string.

Electromagnetic waves are waves that require no material medium for their propagation, and they travel through vacuum (empty space) at an approximate speed of 3.0×10^8 m/s. Electromagnetic waves are governed by Maxwell's electromagnetic equations (laws). They are oscillating electric forces traveling through space which are accompanied by similar oscillating magnetic forces in planes that are perpendicular to each other. Examples of electromagnetic waves are Gamma rays, X-rays, visible and ultraviolet light, microwaves, radio and television waves, and radar waves.

1.4 Production and propagation of waves

5

Electromagnetic waves travel through vacuum and they are usually produced in space due to cosmic radiation. Radio waves and other telecommunication waves are also produced from radio transmitting circuits with an aerial, and other electromagnetic waves like X-rays can also be produced in the laboratory by accelerating electrons with very high voltages. In the following sub-sections, we will concentrate on mechanical waves.

1.4.1 Production of mechanical waves – water waves and waves on a spring/rope

6

Activity
- (i) Get a basin of water
- (ii) Get a stone and tie it with a thread
- (iii) Deep the tied stone into basin of water repeatedly, you will observe some ripples on the water. These ripples are waves that originate from the point where the stone was immersed and spread outwards.

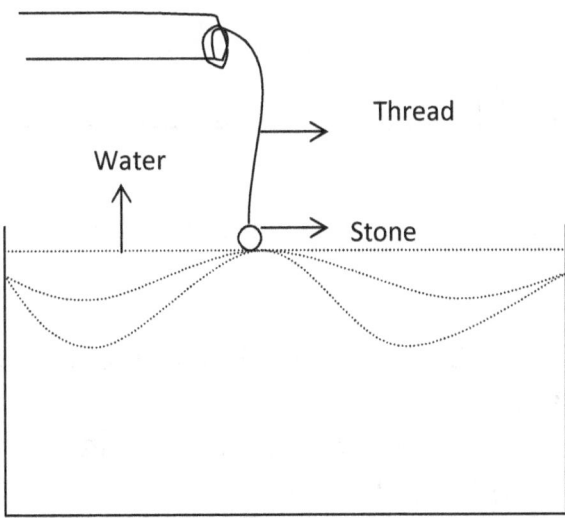

Figure 1.0: A stone dipped into a basin of water

Another way of producing a mechanical wave is by use of rope or spring
Activity B:
   (i)   Get a rope and tie one end of it horizontally to a rigid support
   (ii)   Toss the second end of the rope up and down in a vertical direction
   (iii)   Observe the pulses (waves) that originate from the point of action.

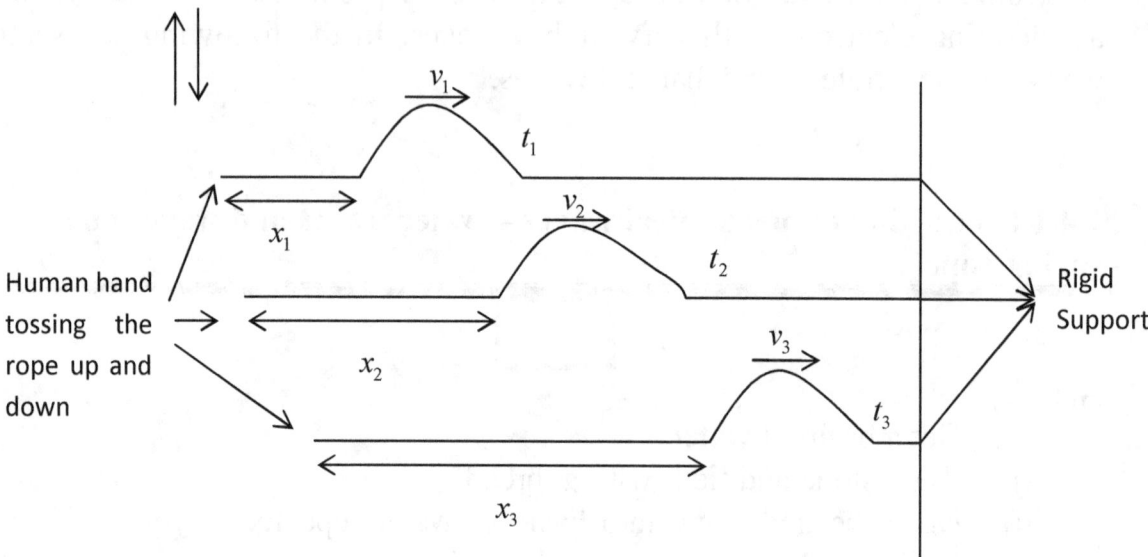

Figure 1.1: A pulse moving along the rope

1.5 Types of wave in terms of mode of propagation

7

When waves are produced, they transfer energy as they propagate or move away from the source of disturbance.

In terms of mode of propagation, waves are divided into progressive (or traveling) waves and standing (or stationary) waves.
Progressive waves are waves that move continuously from one point to another and the vibrations of particles in this type of wave are of the same frequency and amplitude but the phase of vibration changes. Progressive waves are divided into transverse waves and longitudinal waves.

Transverse waves are waves whose mode of propagation is perpendicular to the direction of the vibration of the medium (see figure 1.0 above).
Examples are water waves, waves generated from a plucked spring or rope, and electromagnetic waves.

NOTE: All electromagnetic waves are transverse waves but not all transverse waves are electromagnetic waves.

Longitudinal waves are waves whose mode of propagation is in the same direction (or parallel) to the direction of the vibration of the medium. Examples are sound waves, waves in stretched strings, primary seismic waves, etc.

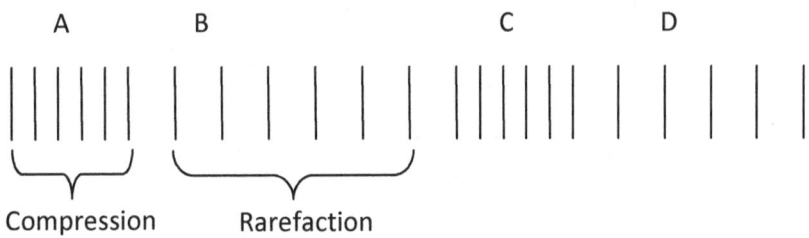

Figure 1.2: Sound wave showing series of compressions and rarefactions

In figure 1.2 above, the vibration of air particles is in the same direction as the propagation of the sound wave.

A common property for progressive waves (both transverse and longitudinal) is that the vibrations of the particles are of the same amplitude and frequency but the phase of vibration changes for different points along the wave.

Stationary waves occur due to superposition of two waves of equal amplitude and frequency traveling in opposite directions along the medium. Stationary waves have the following characteristics or properties:

(i)   They have nodes (where the displacement of the wave is permanently zero).

(ii) Each point along the wave has different amplitude of vibration from neighboring points. The greatest amplitudes are called antinodes.

(iii) At all points between successive nodes, the vibrations are in phase.

(iv) The wavelength, λ, of the wave is twice the distance between successive nodes or successive antinodes. The distance between successive nodes or antinodes is $\lambda/2$ while the distance between a node and a neighboring anti-node is $\lambda/4$.

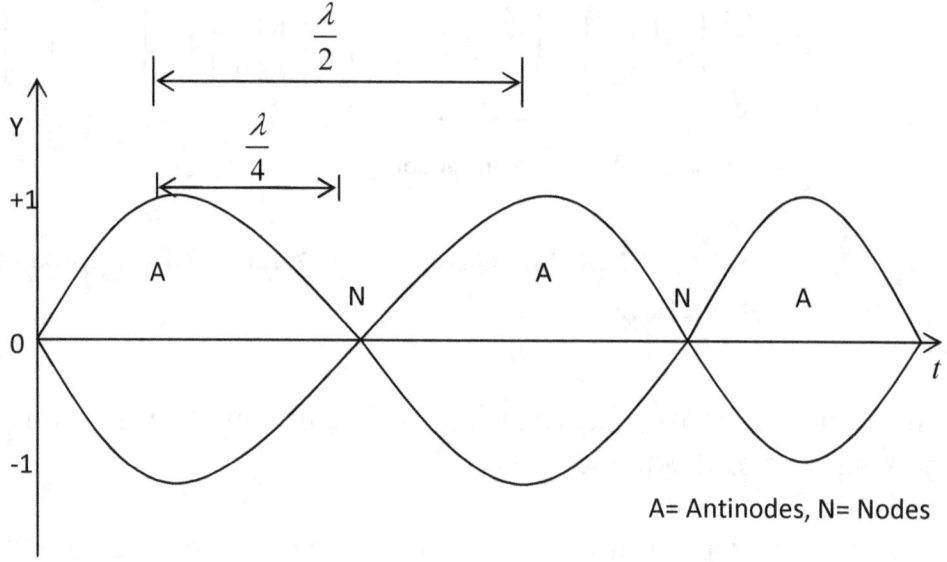

Figure 1.3(a): A sketch of stationary wave

Thus, a stationary wave is one in which some points are permanently at rest (nodes) while antinodes are points between these nodes which are vibrating with varying amplitudes. Examples of stationary waves are waves obtained from a closed pipe and also in an open pipe when the air in them are set into vibration, and waves generated by plucking a string fixed at both ends.

Don't forget!!!

| 10 |

In stationary waves, points between successive nodes or successive antinodes are in phase but vary in amplitude and frequency with each other while in progressive waves, points between successive crest (compression) or successive trough (rarefaction) are out of phase but of the same amplitude and frequency with each other.

1.6.0 Terms used in wave motion – amplitude and wavelength

| 11 |

(i)   Amplitude (A) is the maximum displacement of the particles of the medium from their equilibrium position as the wave passes through them. It is measured in metre (m).

(ii)   Wavelength (λ) is the distance between two successive crests or troughs of transverse waves. It is also the distance between two successive compressions or rarefactions of longitudinal waves.
For stationary waves, the wavelength is twice the distance between two successive nodes or antinodes. Wavelength is also measured in metre (m).

1.6.1 Phase and wave number (k)

| 12 |

If the displacement of a wave is expressed as $y = A_m \sin (kx - wt)$, the phase of the wave is the argument $(kx - wt)$ of the sine which changes with time. Therefore, phase is the maximum angle (in radians) obtained when particles vibrates or undergoes vertical displacement from their mean or rest position.

From the wave function $y = A_m \sin (kx - wt)$ where y = displacement of the wave, and A_m = maximum amplitude,

$$(kx - wt) = \sin^{-1}\left(\frac{y}{A_m}\right)$$

We can therefore define the phase as the sine inverse of the ratio of the

displacement of the vibrating particle at any instant to the maximum amplitude of the vibrating particles.
 NOTE that the 'phase' and 'phase angle' are often used interchangeably and mean the same thing, they are measured in radians.

Wave number (k) is the ratio of angle of particles that have undergone one complete oscillation in one second to the wavelength of the wave. Simply put, the wave number is 2π times the reciprocal of the wavelength of the wave. Wave number is measured in radian per metre (rad/m).

Wave number, $k = \dfrac{2\pi}{\lambda}$

1.6.2 Frequency(f) and Angular frequency (ω)

$$\boxed{13}$$

Frequency (f) of a wave is the number of complete oscillations that a particle makes in one second. It is also the number of complete vibrations or cycles which the wave undergoes in one second. The unit of frequency is Hertz (Hz) or per second (s^{-1}).
Angular frequency (ω) of a wave is 2π times the frequency of the wave; that is, $\omega = 2\pi f$. Angular frequency is measured in radian per second (rad/s).

1.6.3 Period(T) and speed(v)

$$\boxed{14}$$

Period (T) of a wave is the time taken for wave particles to make one complete oscillation. It is also time taken by the wave to travel one wavelength. It is measured in second (s).
Speed (v) of a wave is the distance which the wave travels in one second. It is measured in metre-per-second (m/s).

1.7 Mathematical relationships between the terms used in wave motion

$$\boxed{15}$$

Angular frequency $\omega = 2\pi f$ --- (1)

frequency = reciprocal of the period $= \dfrac{1}{T}$

$$\omega = 2\pi f = 2\pi \times \frac{1}{T} \quad \Rightarrow \omega = \frac{2\pi}{T} \text{ -- (2)}$$

Equation (2) is the relation between the angular frequency and period.

Frequency $f = \dfrac{1}{T}$ but $T = \dfrac{2\pi}{\omega}$, therefore $f = \dfrac{\omega}{2\pi}$ ------------------------------ (3)

$$\text{Speed (v)} = \frac{distance\ travelled\ by\ wave}{time\ taken}$$

$$= \frac{distance\ travelled\ by\ wave\ to\ complete\ 1\ cycle\ (\lambda)}{time\ taken\ to\ complete\ 1\ cycle\ (T)}$$

That is $v = \dfrac{\lambda}{T}$, but $T = \dfrac{2\pi}{\omega}$, therefore $v = \dfrac{\lambda}{2\pi} \times \omega$ ------------------------------ (4)

wave number $k = \dfrac{2\pi}{\lambda}, \quad \Rightarrow \lambda = \dfrac{2\pi}{k}$

Substituting for λ in equation (4) gives: $v = \dfrac{2\pi}{k} \times \dfrac{1}{2\pi} \times \omega = \dfrac{\omega}{k}$

$$v = \frac{\omega}{k} = \frac{\lambda}{T} = f\lambda \text{ -- (5)}$$

Equations $(1) - (5)$ are very important in wave applications.

1.8 Graphical representation of wave motions

16

The displacement, y, of a vibrating particle in a medium in which a wave passes can be represented using sine or cosine curves

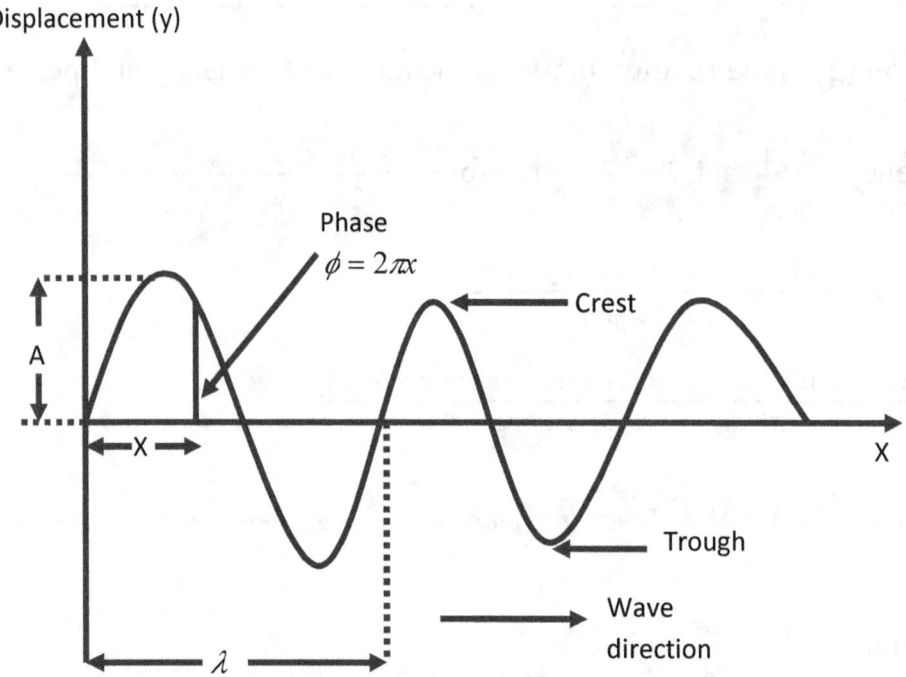

Figure 1.3(b): Graphical representation of progressive waves

Specifically, the above wave profile is a sine or sinusoidal wave form that represents the vibration of the particles of the medium as the wave passes. From the diagram above, the crest and trough are the characteristics of transverse waves while compression and rare-faction are the characteristics of longitudinal waves. Transverse and longitudinal waves can be represented as shown above.

1.9 Mathematical representation of wave motions (Equations of waves)

The displacement of a progressive wave at a particular point and time interval along the x-axis are expressed as follows.

In sine form: $y(x, t) = y_m \sin(\omega t \pm \phi) \Rightarrow y = A_m \sin(\omega t \pm \phi)$ ------------ (6)

In cosine form: $y(x, t) = y_m \cos(\omega t \pm \phi) \Rightarrow y = A_m \cos(\omega t \pm \phi)$ ------------ (7)

Where $y(x, t)$ = Displacement of the particles; $y_m = A_m$ = amplitude; $(\omega t \pm \phi)$ = phase; ω = angular frequency; and t = time.

18

The plus-or-minus ($\pm$) signs in wave equations expressed in plan 17 indicates the direction of the wave. If the wave travels from left to-right, we use the minus (-) sign, but if the wave travels from right-to-left, we use the plus (+) sign.

In the following sections, we will assume that the wave motion travels from left-to-right and sinusoidal wave forms will be considered.

1.9.1 Other forms of Wave Equations

19

If we assume that the wave motions are sinusoidal and travel from left-to-right, then we have

$y = A_m \sin (\omega t - \phi)$

but ϕ = phase angle = kx

Therefore $y = A_m \sin (\omega t - kx) \Leftrightarrow y = A_m \sin (kx - \omega t)$ ----------------------- (8)

Next, putting k = wave number = $\dfrac{2\pi}{\lambda}$, and ω = angular frequency = $2\pi f$, we have

$$y = A_m \sin \left(2\pi f t - \frac{2\pi x}{\lambda} \right)$$ -- (9)

but speed of the wave, $v = f\lambda \Rightarrow f = \dfrac{v}{\lambda}$

substituting this into equation (9) gives:

$$y = A_m \sin \left(\frac{2\pi v t}{\lambda} - \frac{2\pi x}{\lambda} \right) = A_m \sin \frac{2\pi}{\lambda} (vt - x)$$ ------------------------------------- (10)

From equation (8):

$$y = A_m \sin 2\pi \left(ft - \frac{x}{\lambda} \right), \qquad but \ f = \frac{1}{T}$$

Therefore $y = A_m \sin 2\pi \left(\dfrac{t}{T} - \dfrac{x}{\lambda} \right)$ --- (11)

Remarks!!!

Equation (7) involves angular frequency (ω) and wave number (k), equation (8) involves frequency (f) and wave length (λ), equation (9) involves the speed (v) and wave length, and equation (10) involves the period (T) and wave length.

Other parameters remain constant. Moreover, if you are given a problem involving equation of wave, study the question carefully to know the parameters given and the ones you are looking for. Let's play with some examples!

Example 1

A plane progressive wave is represented by the equation $y = 0.1\sin(200\pi t - 20\pi x/17)$, where y is the displacement in millimeters, t is in seconds and x is the distance from a fixed origin O in metres (m). Find
  (i)   the frequency of the wave,
  (ii) the wave length,
  (iii) the wave velocity, and
  (iv) the phase difference in radians between a point 0.25m from O and a point 1.10 from O.

Solution:

In questions like this, it is beautiful and elegant to compare the given wave equation with any of the standard wave equations just discussed.

Given: $y = 0.1\sin(200\pi t - 20\pi x/17)$
Standard: $y = A_m\sin(\omega t - kx)$

Comparing the 2 gives:

$A_m = 0.1\,\text{mm} = 0.1\times10^{-3}\,\text{m}$

$\omega = 200\,\pi$

$k = \dfrac{20\,\pi}{17}$

Now, try and see if you can finish the problem on your own!

Could you do that? Look in here!

23

(i) $\omega = 200\pi$, $\quad f = \dfrac{\omega}{2\pi} = \dfrac{200\pi}{2\pi} = 100\,Hz$

That is the frequency of the wave.

(ii) $k = \dfrac{20\pi}{17}$, $\quad but \quad k = \dfrac{2\pi}{\lambda} \Rightarrow \lambda = \dfrac{2\pi}{k} = \dfrac{2\pi}{\frac{20\pi}{17}} = \dfrac{2\pi}{20\pi}\times\dfrac{17}{1} = \dfrac{17}{10} = 1.7\,m$

That is the wavelength.

(iii) Speed, $v = f\lambda = 100\times1.7 = 170\,m/s$

(iv) Phase angle, $\phi = \dfrac{2\pi x}{\lambda}$

at x = 0.25m; $\quad \phi_{\text{at }0.25m} = \dfrac{2\pi}{1.7}\times0.25 = 0.92$

at x = 1.10; $\quad \phi_{\text{at }1.10m} = \dfrac{2\pi\times1.10}{1.7} = 4.07$

Therefore phase difference in question is:

$\phi_{\text{at }1.10m}$ - $\phi_{\text{at }0.25m}$ $\quad = 4.07 - 0.92 \quad = 3.15$

Hey, look at this JAMB Question!

24

A traveling wave moving from left to right has amplitude of 0.15m, a frequency of 550Hz and a wavelength of 0.01m. The equation describing the wave is

(A) $y = 0.15\sin 200\pi(x - 5.5t)$

(B) $y = 0.15\sin \pi(0.01x - 5.5t)$

(C) $y = 0.15\sin 5.5\pi(x - 200t)$

(D) $y = 0.15 \sin \pi(550x - 0.01t)$

[JAMB 2000]

Try the problem first, and then check with the solution below:

Solution:

Amplitude $A_m = 0.15$m, frequency f = 550Hz, and wave length $\lambda = 0.01$m

The standard wave equation is: $y = A_m \sin(\omega t - kx)$

where $\omega = 2\pi f$, and $k = \dfrac{2\pi}{\lambda}$

$$\text{Therefore } y = A_m \sin\left(2\pi f t - \frac{2\pi x}{\lambda}\right)$$

$$y = 0.15 \sin\left(2\pi \times 550t - \frac{2\pi x}{0.01}\right)$$

$$y = 0.15 \sin 2\pi\left(550t - \frac{x}{0.01}\right)$$

$$y = 0.15 \sin 2\pi (550t - 100x)$$

$$y = 0.15 \sin 200\pi (5.5t - x)$$

So, option (A) is the correct!

Another JAMB Question

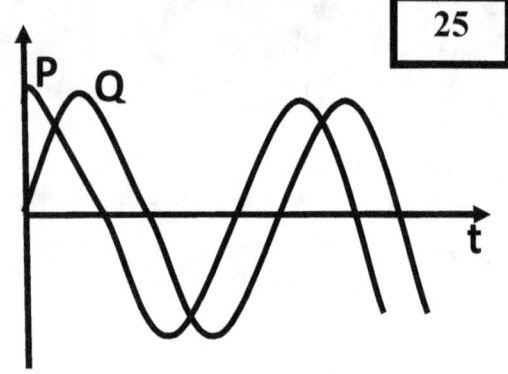

The phase difference between wave P and Q in the diagram above is

(A) 2π (B) π (C) $\dfrac{\pi}{4}$ (D) $\dfrac{\pi}{2}$

[JAMB 2004]

If you got $\frac{\pi}{2}$, then you are right!

Solution:
It can be seen from the diagram that the distance between the crest of wave P and the next immediate crest of wave Q is λ/4 (That is, a quarter of the wavelength, see diagram below).

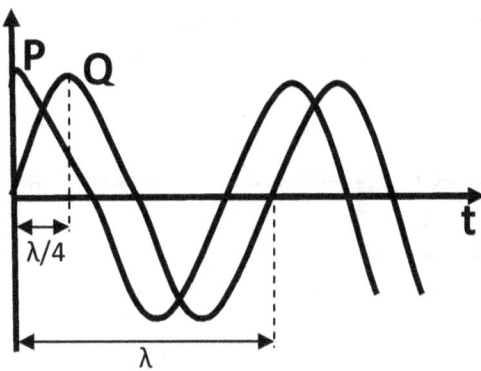

Therefore the phase difference between the waves is = $\phi = \dfrac{2\pi x}{\lambda} = \dfrac{2\pi}{\lambda} \cdot \dfrac{\lambda}{4} = \dfrac{\pi}{2}$

And so option (D) is correct!

Next example!

26

A wave is represented by the equation y = 0.2 sin 0.4π (x − 60t), where all distances are measured in cm and time in seconds. Calculate the speed of the wave. What is the displacement of the wave at a distance of 15m from the origin at time 0.02s?

Solution:

Given wave equation: y = 0.2 sin 0.4π (x − 60t)
                        ⇒ y = 0.2 sin (0.4 π x - 24 π t)
Standard wave equation: $y = A_m \sin (kx - \omega t)$

Comparing the 2 gives:
A_m = 0.2cm = 0.002m,

$k = 0.4\pi$, and
$\omega = 24\pi$

Now, $\lambda = \dfrac{2\pi}{k} = \dfrac{2\pi}{0.4\pi} = 5cm = 0.05m$

and $\quad f = \dfrac{\omega}{2\pi} = \dfrac{24\pi}{2\pi} = 12Hz$

Therefore, Speed V = $f\lambda$ = 0.05 × 12 = 0.6m/s OR 60cm/s.

Did you get that?

$$\boxed{27}$$

Now, let's find the displacement, y, at x = 15cm and t = 0.02s
y = 0.2 sin (0.4πx - 24πt)
  = 0.2 sin (0.4π × 15 - 24π × 0.02)
  = 0.2 sin (18.85 – 1.507)
  = 0.2 sin (17.343)
  = 0.2 × 0.298 = 0.059cm
  $\cong$ 0.06cm = 6×10^{-4}m

Do this last and quick one on your own!

$$\boxed{28}$$

What is the frequency of a radio wave of wavelength 150m if the velocity of radio wave in free space is 3.0×10^8m/s?

Solution: λ = 150m, v = 3.0×10^8, f = ?
Using the relation, v = $f\lambda$

$f = \dfrac{v}{\lambda} = \dfrac{3 \times 10^8}{150} = 2 \times 10^6 \ Hz$

You sure got it! So, we take off to the next agenda.

Waves (whether mechanical or electromagnetic waves) exhibit the following properties; reflection, refraction, diffraction, interference, and polarization.

JUST AN EMPHASIS: All electromagnetic waves are transverse waves but not all transverse waves are electromagnetic waves.
Transverse waves can be plane-polarized while longitudinal waves can not be plane-polarized. This is the striking difference between transverse and longitudinal waves.

1.10.1

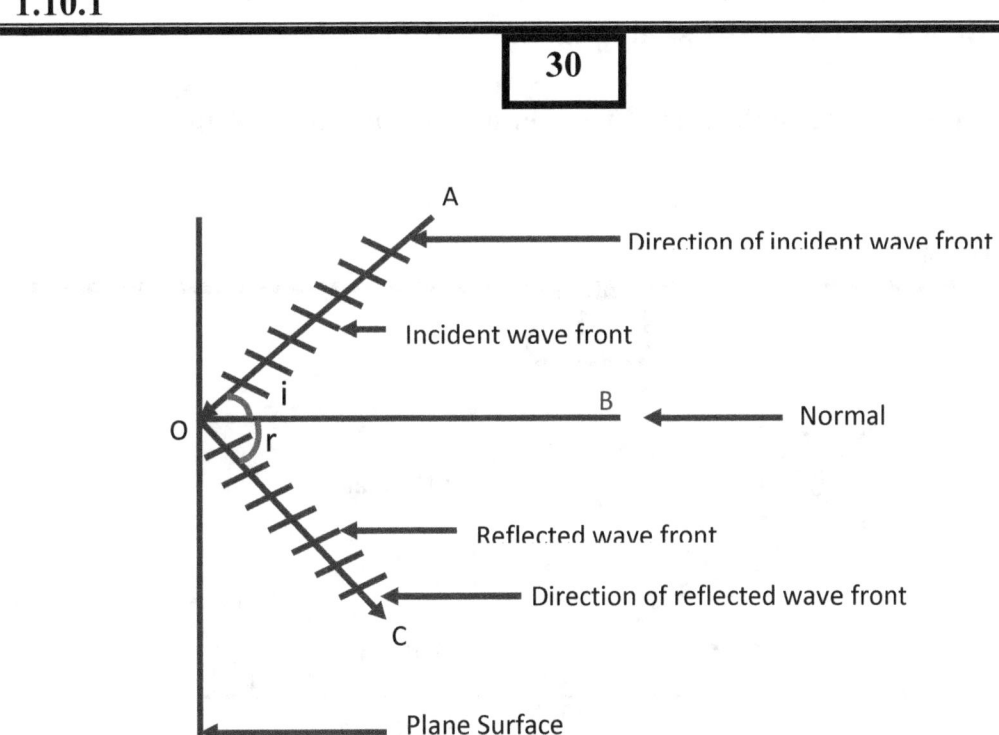

Figure 1.4: Reflection of waves

Waves are said to be reflected when they bounce off surfaces like mirrors. As shown in figure 1.4, the wave forms of the reflected waves depend on the nature and shape of the surface.

Explanation:

Using a water tank, for instance, when the vibrator is switched on, waves are generated, spreading outwards and when they strike a plane surface, reflected wave fronts are obtained in opposite direction of the incident wave fronts. The angle which the incident wave makes with the normal to the surface is called the angle of incidence, i, while the angle which the reflected wave makes with the normal to the surface is called the angle of reflection, r. For waves to reflect from any surface, the angle of incidence must be equal to the angle of reflection. This is a reflection law.

The laws of reflection state that:
1. **The incident ray, the reflected ray, and the normal at the point of incidence all lie on the same plane.**

2. **The angle of reflection is always equal to the angle of incidence.**

1.10.2 Refraction of waves.

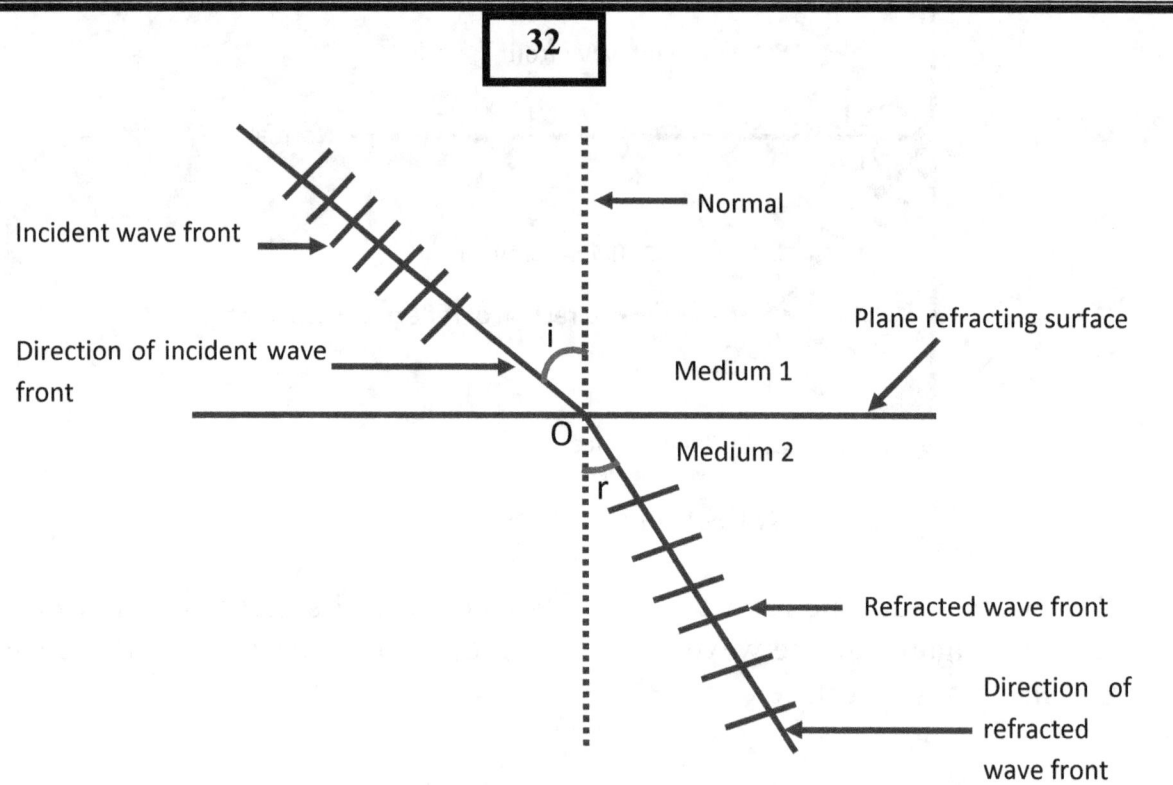

Figure 1.5: Refraction of waves

Refraction of waves is the change in the speed and direction of the waves as they pass the boundary between two media of different densities, as shown in figure 1.5.

Explanation.

33

When a plane light wave front strikes the surface of transparent materials like triangular glass prisms or rectangular glass blocks, the incident wave front bends as it passes through the material. This is because the light wave front has traveled from a medium of less optical density to one of more optical density.

The angle between the incident wave and the normal to the boundary surface is called the angle of incidence, while the angle between the normal and the refracted wave is called angle of refraction.

Materials that refract waves are often characterized by a quantity called the refractive index. The refractive index of a material/medium (e.g. glass) with respect to another material/medium (e.g air), is given as:

$$_a\eta_g = \frac{velocity\ of\ incident\ wave}{velocity\ of\ refracted\ wave} = \frac{v_1}{v_2}$$

If the frequency of the wave is constant in both media, $v_1 = f\lambda_1$, and $v_2 = f\lambda_2$, then:

$$_a\eta_g = \frac{v_1}{v_2} = \frac{f\lambda_1}{f\lambda_2} = \frac{\lambda_1}{\lambda_2}, \quad that\ is \quad \frac{wavelength\ of\ the\ waves\ in\ medium\ 1\ (e.g.\ air)}{wavelength\ of\ the\ wavesf\ in\ medium\ 2\ (e.g.\ glass)}$$

Also, $\quad _a\eta_g = \frac{Sine\ of\ angle\ of\ incidence}{Sine\ of\ angle\ of\ refraction} = \frac{Sin\ i}{Sin\ r}$

This is a refraction law, also called Snell's law of refraction.

In summary: $\quad _a\eta_g = \frac{v_1}{v_2} = \frac{\lambda_1}{\lambda_2} = \frac{Sin\ i}{Sin\ r}$

The laws of refraction state that:

1. **The incident ray, the refracted ray, and the normal at the point of incidence all lie on the same plane.**

2. **The ratio of the sine of the angle of incidence to the sine of the angle of refraction is a constant.**

   That is $\dfrac{Sin\,i}{Sin\,r} = \eta$**, where** η **is the constant called the refractive index.**

1.10.3 Diffraction of waves

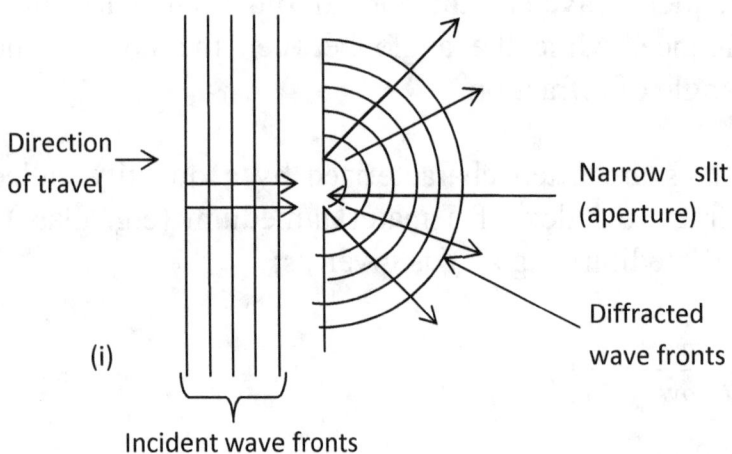

(i)

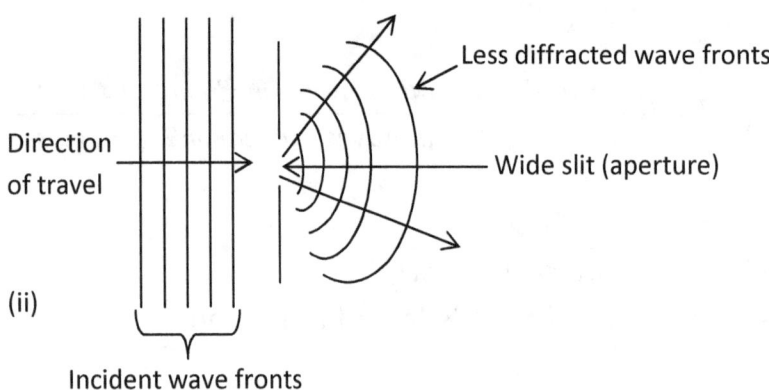

(ii)

Figure 1.6: Diffraction of waves from narrow and wide apertures

Diffraction of waves is the spreading of waves when they pass through apertures or around obstacles. The wave forms or patterns of the Diffracted waves depends on the size (width) of the aperture and on the magnitude of the wavelength of the incident wave fronts, that is, the smaller the width of the aperture in relation to the wavelength, the greater is the spreading or diffraction of the waves and vise versa.

Explanation:

35

Diffraction of waves is also the ability of waves to bend around obstacles in their paths. When progressive waves pass through narrow openings or move around obstacles, they tend to spread out in different directions. This occurs when the wavelength of the waves is longer than the size or width of the obstacle or aperture.

1.10.4 Interference of waves

36

If two or more waves overlap, the resultant displacement is the sum of the individual displacements. The overlapping of waves is also technically called interference. Stationary or standing waves are obtained when two waves of equal amplitude and frequency (or wavelength) traveling in opposite directions combine together.

When two waves, say A and B, superimpose on each other to produce a maximum disturbance at a particular point, a constructive interference is obtained but when the two waves combine or superimpose and produce a minimum or zero disturbance at a particular point, a destructive interference is obtained.

For constructive interference to occur, the path difference between the two waves must be zero or multiple values of the wavelength, that is, path difference = 0 or λ, or 2λ or 3λ . . . . .

For destructive interference to occur, the path difference between the two waves must be fractional values of the wavelengths, that is, path difference = $\frac{\lambda}{2}$ or $\frac{3}{2}\lambda$ or $\frac{5}{2}\lambda$ or $\frac{7}{2}\lambda$ . . . . .

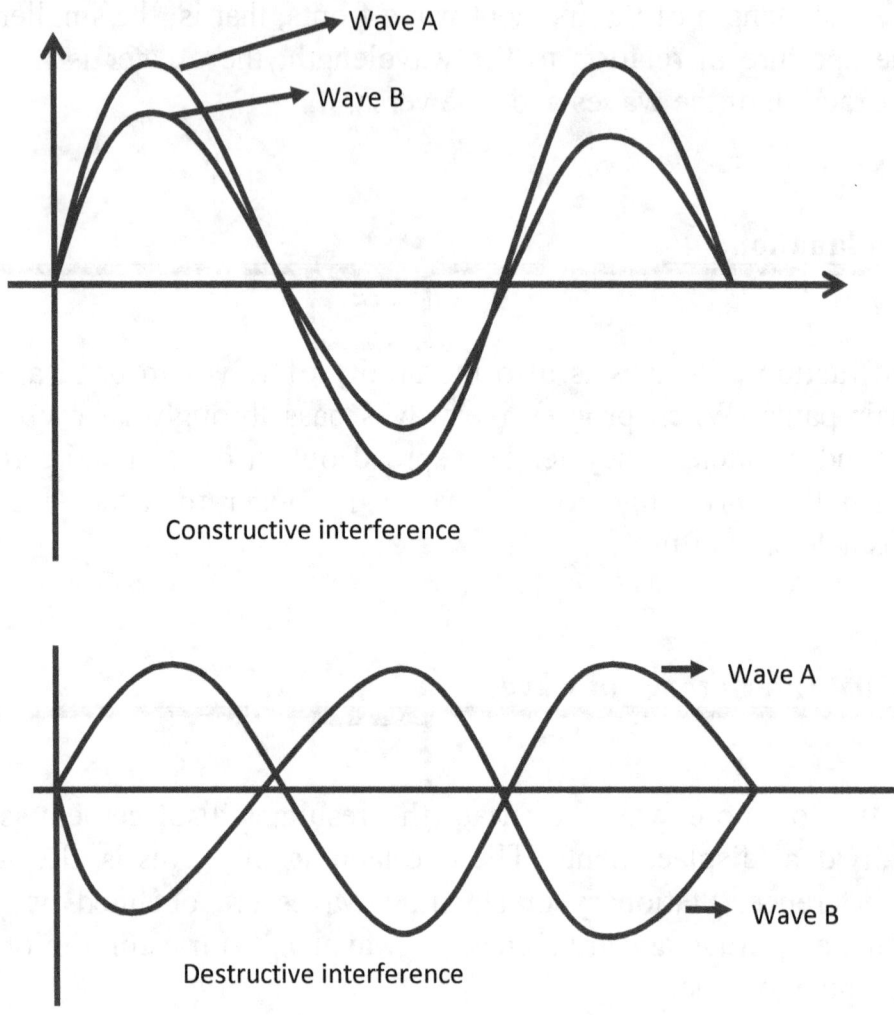

Figure 1.7: Constructive and Destruction interference of two waves A and B

NOTE: For two waves to interfere, they must be produced from coherent wave sources. Coherent wave sources are wave sources which produce waves that have the same frequency or wavelength and amplitude and the two waves are always in phase with each other or they have a constant phase difference.
Interference can be demonstrated in the ripple tank by using the two point sources. Figure 1.8 below shows such an interference pattern.

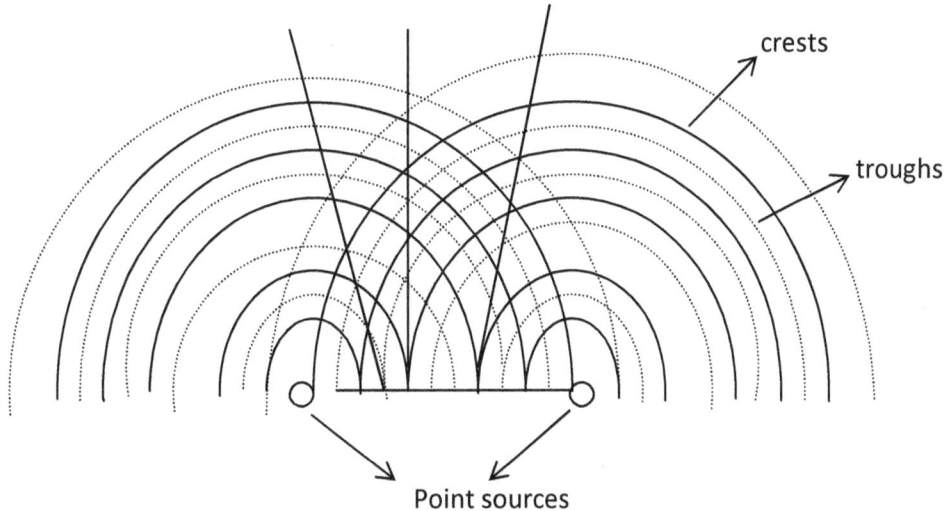

Figure 1.8: Two source interference of circular waves in a ripple tank

1.10.5 Polarization of waves

Polarization of waves is the process of making waves that have a diffused mode of propagation (waves whose direction and mode of vibration are dispersed randomly) to vibrate only in one plane or to have regular pattern (or shapes). Our concern here is light waves; polarization of light waves is the process of making unpolarised light wave to have regular patterns in their mode of propagation. Unpolarised light wave is light wave that have no definite wavelength. Polarization of waves is achieved by use of polarisers and such waves are said to be plane-polarized. Waves are plane-polarized if their vibrations occur only in one plane. Therefore polarization is a phenomenon in which waves whose vibrations are only in one plane are produced.

NOTE: Only transverse waves (example electromagnetic waves and some mechanical waves such as waves generated on a plucked spring or rope) can be plane-polarized while longitudinal waves such as sound waves cannot be plane-polarized.

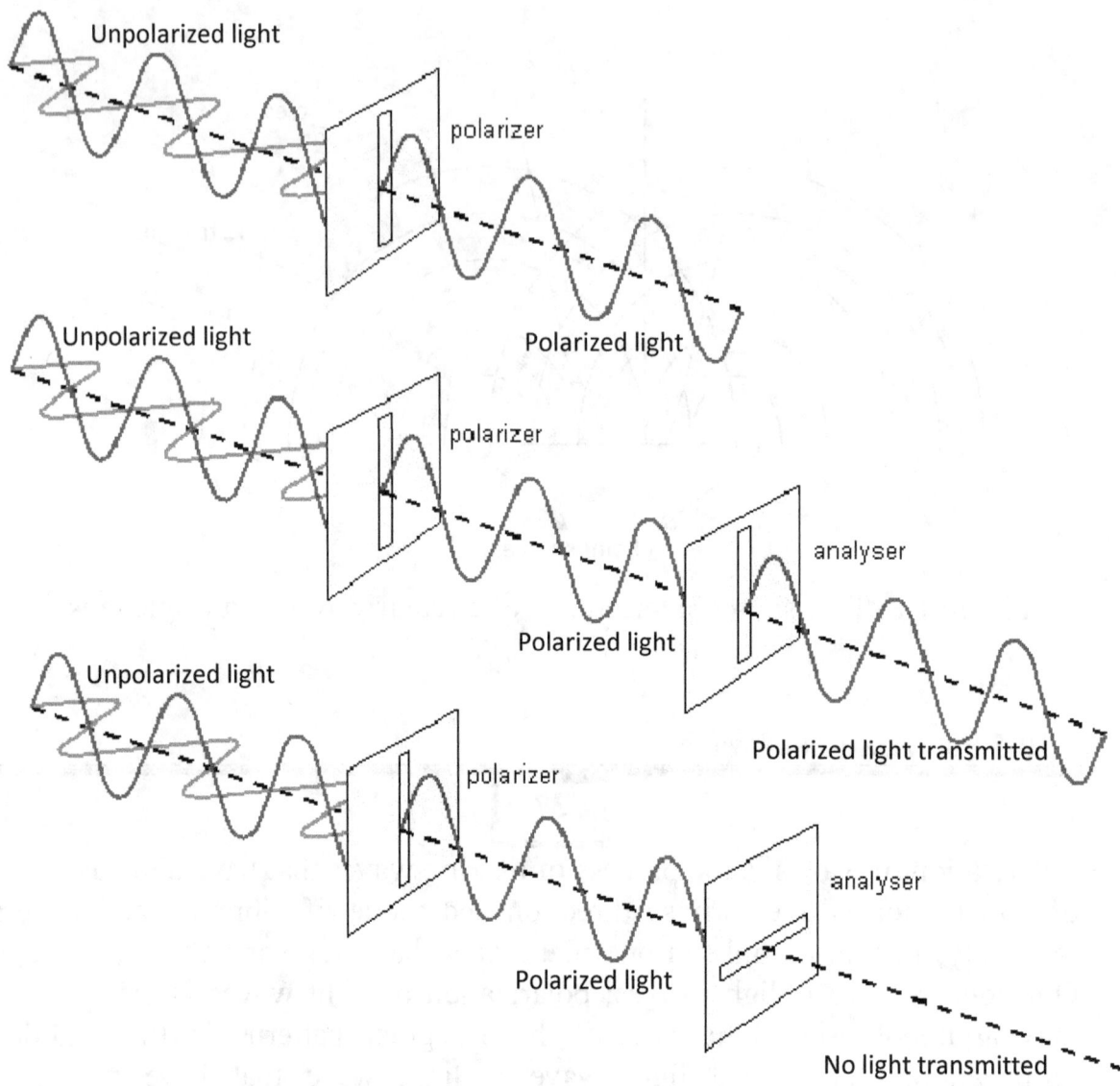

Figure 1.9: Polarization of light waves (http://www.schoolphysics.co.uk)

1.11 Polarizers (polaroids) and analyzers

Polarisers are filters that only allow wave vibrations to pass through a particular direction. Polaroids or polarisers are also slits that makes unpolarised light wave to be plane-polarized. They have the ability of making waves that have diffused direction to have a regular pattern, as shown in figure 1.9.

Examples of polarizers are Quartz Polaroid and Tourmaline crystal.

Analyzers are also polarizers that are used to check if waves incident on them are polarized or not; if the waves are polarized, they pass through analyzers that have their transmission axis aligned in the plane of polarization, but they do not pass through analyzers that have their transmission axis perpendicular to the plane of polarization.

Applications of polarisers (polaroids) or polarized light

Polarisers are used
    (1) In sunglasses to reduce the intensity of incident light.
    (2) On "head-edge" of vehicle wind screen to eliminate reflected light or glare from the road.
    (3) In the production and visualization of three-dimensional films
    (4) In determination of concentration of sugar solution
    (5) In production of plane-polarized light from unpolarized light as in polarized cameras.

There we go! We've come to the end of the class. It's time to lay your hands on the exercises and see how much you understood. Goodluck!

Exercises

1. A progressive wave is given as $y = 0.5 \sin 2\pi (40t - 20x)$, find the
(i) wavelength   (ii) frequency   (iii) period   (iv) speed   (v) displacement at $x = 5cm$ and $t = 0.02s$, of the wave. y is in meter and t is in second.

2. A wave is represented by the equation $y = 0.2 \sin (0.4\pi x - 24\pi t)$, where all distances are measured in centimeters and time in second. Calculate the speed of the wave.

3. Radio waves travel at a velocity of $3.0 \times 10^8 m/s$ in air. Calculate the wavelength in air of radio waves when transmitted at a frequency of 150MHz.

4. A periodic pulse travels a distance of 20m in 1.00s. If its frequency is $2.0 \times 10^3 Hz$. calculate the wavelength of the pulse.

5. A straight vibrator causes water ripples to travel across the surface of a shallow tank. The waves travel a distance of 33cm in 1.5s, and the distance between two successive wave crests is 4.0cm. Calculate the frequency of the vibrator.

6. A wave is represented by the equation; $y = 2\sin \pi (0.5x - 200t)$, where all distances are measured in centimeter and time in second. Calculate (i) frequency (ii) wavelength (iii) speed of the wave.

7. Light of wavelength 6.0×10^{-7} m travels in air at a speed of 3.0×10^8 m/s.
(i) Calculate the frequency of this light.
(ii) State the effect, if any, on the frequency as the light enters the glass from air.
(Cambridge, 2011)

8.

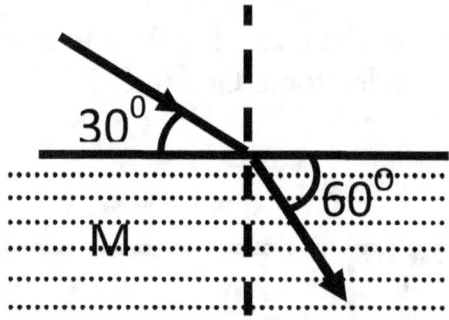

The refractive index of the medium M in the diagram above is

A. $\frac{2}{\sqrt{3}}$ B. $\frac{1}{\sqrt{3}}$ C. $2\sqrt{3}$ D. $\sqrt{3}$

(JAMB, 2004)

9.

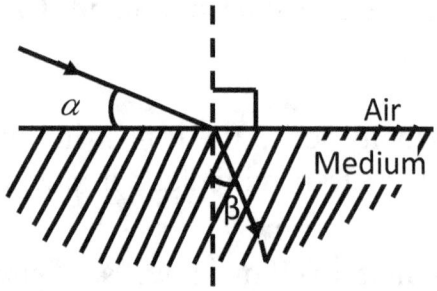

If a ray traveling in air is incident on a transparent medium as shown in the diagram, the refractive index of the medium is given as

A. $\dfrac{\cos \alpha}{\sin \beta}$ B. $\dfrac{\sin \alpha}{\sin \beta}$ C. $\dfrac{\cos \beta}{\sin \alpha}$ D. $\dfrac{\sin \beta}{\sin \alpha}$

(JAMB, 2001)

10. Vibrations in a stretched spring cannot be polarized because they are
A. Longitudinal waves
B. Mechanical waves
C. Stationery waves
D. Transverse waves
(JAMB, 2002)

11. If a light wave has a wavelength of 500nm in air, what is the frequency of the wave?
A. 3.0×10^{14} Hz
B. 6.0×10^{14} Hz
C. 6.0×10^{12} Hz
D. 2.5×10^{14} Hz
[$c = 3.0 \times 10^{8}$ m/s]
(JAMB, 2009)

12. A ray of light enters a glass block at an angle of incidence i, producing an angle of refraction r in the glass.

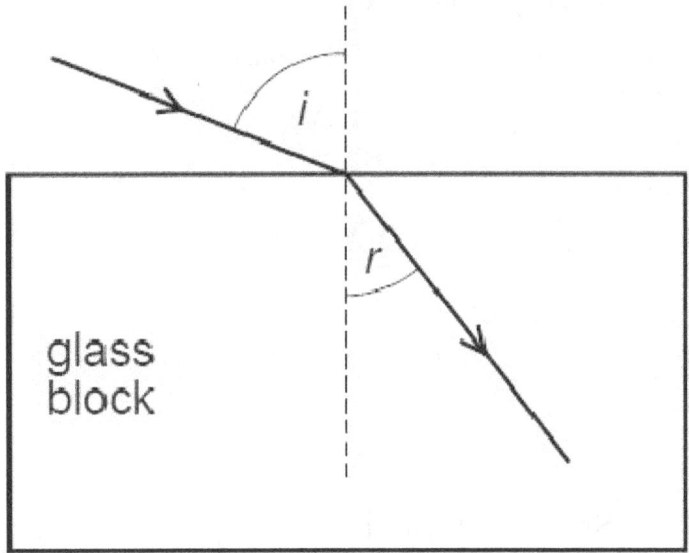

Several different values of i and r are measured, and a graph is drawn of sin i against sin r. Which graph is correct?

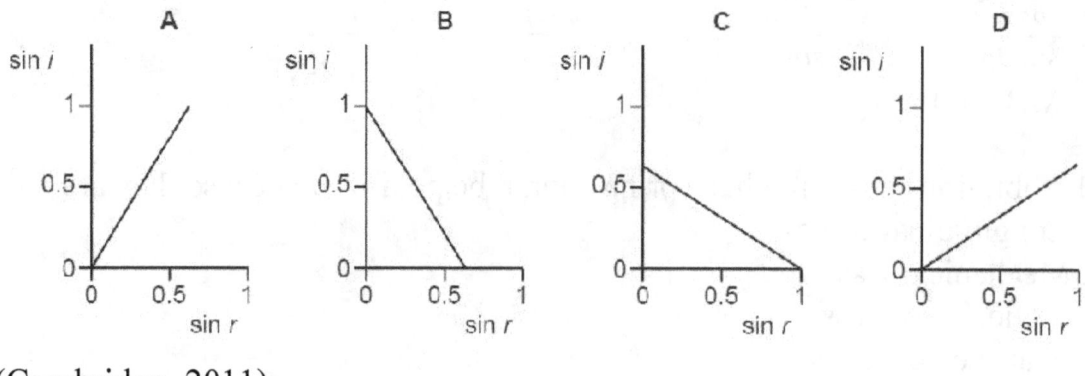

(Cambridge, 2011)

Solution to Exercises

| 41 |

1. (i) 0.05m (ii) 40Hz (iii) 0.025s (iv) 2m/s (v) -0.475m
2. 60cm/s OR 0.6m/s
3. 2m
4. 0.01m
5. 5.5Hz
6. (i) 100Hz (ii) 4cm OR 0.04m (iii) 400cm/s OR 4m/s
7. (i) 5×10^{14} Hz   (ii) the frequency does not change, but the speed and wavelength will both decrease.

8. Option D is correct (Hint & Note: the angle of incidence is not $30°$ but $90 - 30 = 60°$, similarly the angle of refraction is not $60°$ but $90 - 60 = 30°$).

9. Option A is correct (Hint & Note: the angle of incidence is not α but $90 - \alpha$, and $\sin(90 - \alpha)$ is the same as $\cos \alpha$. The angle of refraction is truly β.

10. Option A is correct
11. Option B is correct
12. Option A is correct; from Snell's law, sin i is directly proportional to sin r, and η is the constant of proportionality which is the slope of the graph in this case.

www.ingramcontent.com/pod-product-compliance
Lightning Source LLC
Chambersburg PA
CBHW081247170526
45165CB00009B/3232